ANIMAL RHYME TIME!

RHYME TIME WITH PARROTS!

BY JONAS EDWARDS

Gareth Stevens
PUBLISHING

Please visit our website, www.garethstevens.com. For a free color catalog of all our high-quality books, call toll free 1-800-542-2595 or fax 1-877-542-2596.

Cataloging-in-Publication Data

Names: Edwards, Jonas.
Title: Rhyme time with parrots! / Jonas Edwards.
Description: New York : Gareth Stevens Publishing, 2021. | Series: Animal rhyme time! | Includes glossary and index.
Identifiers: ISBN 9781538255988 (pbk.) | ISBN 9781538256008 (library bound) | ISBN 9781538255995 (6 pack)
Subjects: LCSH: Parrots–Juvenile literature.
Classification: LCC QL696.P7 E39 2021 | DDC 598.7′1–dc23

First Edition

Published in 2021 by
Gareth Stevens Publishing
111 East 14th Street, Suite 349
New York, NY 10003

Editor: Kate Mikoley

Photo credits: Cover, p. 1 spotters/Shutterstock.com; cover, pp. 3–24 (music notes) StockSmartStart/ Shutterstock.com; p. 5 Andreas Speich / EyeEm/Getty Images; p. 7 Peter Nordbaek Hansen/ Shutterstock.com; p. 9 SpeedPhoto/ iStock / Getty Images Plus; p. 11 Nana_Hana/Shutterstock.com; p. 13 mayskyphoto/ iStock / Getty Images Plus; p. 15 Enrique Aguirre Aves/ Oxford Scientific / Getty Images Plus; p. 17 Tier Und Naturfotografie J und C Sohns/ The Image Bank / Getty Images Plus; p. 19 Isabel Pavia/Moment/Getty Images; p. 21 Stephen M. Reed/Moment/Getty Images.

Printed in the United States of America

CPSIA compliance information: Batch #CS20GS: For further information contact Gareth Stevens, New York, New York at 1-800-542-2595.

CONTENTS

Boldface words appear in the glossary.

Talking Birds

Have you ever seen a bird that can talk before?

Parrots are known for doing that—and much more!

They come in bright colors, like red, green, and yellow.

A pet parrot's first word is often "hello."

Parrot Living

In the wild, most parrots live where it's warm.

Although some live in mountains, where snow can form.

Parrots stay together in big groups called flocks.

In areas without trees, some make their homes in the rocks.

Beak Eats

Parrots eat mainly plants, nuts, and seeds.

Some also eat bugs to meet their **nourishment** needs.

A parrot's strong **beak** can open up foods.

That's how they eat the hard nuts and seeds their **diet** includes.

Funny Feet

Now you know what parrots like
to eat.
Another thing all have in common is
their zygodactyl feet.
This means two toes face behind
and two face the front.
This helps them hold food after a
seed or fruit hunt.

Chatty Parrots

You already know that parrots can talk.
But did you know they also sing
and squawk?
One parrot knew how to say more
than 1,700 words.
That's quite a lot—these animals sure
are chatty birds!

A talking parrot isn't really having
a **conversation**.
It's really just copying a
person's **declaration**.
African gray parrots like to talk but
don't know what they're saying.
Mostly they just **mimic** what they hear,
like a funny kind of playing.

Many Shapes and Sizes

Baby parrots don't have many feathers until they get older.

They come in all different colors, mostly bright, but some duller or bolder.

Parrots also come in many sizes and weights.

But most male parrots look a lot like their female **mates**!

Pet Parrots

Many people have a parrot as a pet. If you want one, it's important to think before you get. Some parrots live for 60 years or more. They're messy and loud, so taking care of them can be a **chore**.

Let's See a Parrot!

Now you know how much parrots rule.
Perhaps you'll see one at the zoo on a trip with school.
What color bird do you think it will be?
Maybe it will be yellow and blue. You'll just have to see!

GLOSSARY

beak: a part of the mouth that sticks out on some animals and is used to tear food

chore: task

conversation: talk

declaration: something that is said out loud or made known

diet: the food that an animal eats

mate: one of two animals that come together to produce babies

mimic: to copy

nourishment: food and other things needed to be healthy and grow strong

FOR MORE INFORMATION

BOOKS

Moses, Brian. *Animal Poems.* New York, NY: Windmill Books, 2018.

Randolph, Joanne. *Poems About Animals.* New York, NY: Windmill Books, 2019.

Statts, Lee. *Parrots.* Minneapolis, MN: Abdo Zoom, 2018.

WEBSITES

Parrots
www.dkfindout.com/us/animals-and-nature/birds/parrots/
Head to this interactive page to learn even more about parrots.

Parrots
www.nationalgeographic.com/animals/birds/group/parrots/
Find out more fun facts about parrots here!

INDEX